SPACE CENTERS

SMITHMARK

This edition first published in 1992 by SMITHMARK
Publishers Inc., 112 Madison Avenue,
New York, New York 10016

ISBN 0-8317-0507-8

Printed and bound in Spain

Writer: Edward Hymoff
Designer: Ann-Louise Lipman
Design Concept: Lesley Ehlers
Editor: Sara Colacurto
Production: Valerie Zars
Photo Researcher: Edward Douglas
Assistant Photo Researcher: Robert V. Hale
Editorial Assistant: Carol Raguso

Title page: An early December sunset
starkly outlines the *Apollo 17* booster
and spacecraft sitting on the Kennedy
Space Center launch pad at Cape
Canaveral, Florida, as it is prepared
for the final visit to the moon by U.S.
astronauts on December 7–19, 1972.
Opposite: An assemblage of rocket
boosters, a "must see" tourist stop for
visitors from all parts of the world, is on
display at the Kennedy Space Center.

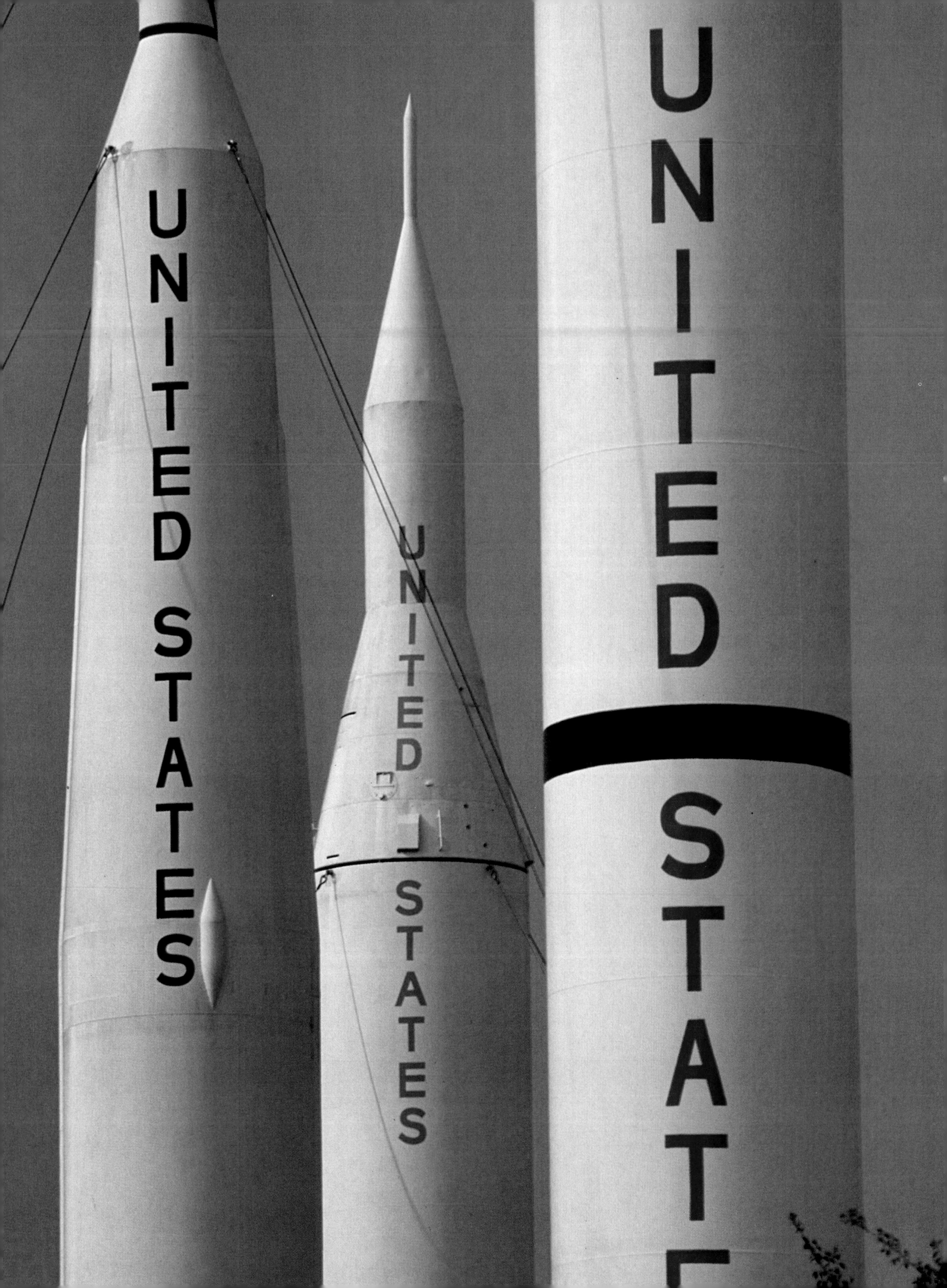
UNITED STATES
UNITED STATES
UNITED STAT

O ne of humankind's greatest accomplishments has been the exploration of space, a result of the combined efforts of skilled individuals highly trained in the diverse fields of science, engineering, and technology. Some 4,500 years ago the same might have been written about the building of Egypt's ancient pyramids, its Sphinx, and its huge temples, which were carved out of solid rock—edifices that, although worn and weather-beaten with time, continue to amaze and inspire us.

On the moon, exposed to the silent and destructive elements of space only since 1969, are amazing experimental space-age operating systems and other equipment left behind by *Apollo* astronauts that transmit crucial data to Earth. On Mars and Venus, unmanned satellites sent by the United States and the Soviet Union on exploratory missions have been landed, supplying information about the atmosphere and surface of both these planets. In an effort to learn more about space, near and far, the United States continues to launch unmanned satellites to the outer planets of our solar system: Saturn, Pluto, Jupiter, Mercury, Uranus, and Neptune.

But perhaps the greatest achievements in space exploration are yet to come. Many nations will soon participate in building the manned space station *Freedom,* a permanent observation platform that will orbit Earth and provide important data about worldwide agricultural and environmental problems. Also planned is a return to the moon to establish a permanent

Preceding page: Rocket boosters at Kennedy stand against the same sky that their predecessors traversed on their race to conquer space. *This page, top to bottom:* Spacecraft that were never launched are also on display at the Kennedy Space Center. This *Apollo* lunar module was used for ground testing and astronaut training during preparations for lunar missions. Telecommunications antennae appear to tower above the Kennedy headquarters building as tourists walk past on their way to the space center's many historic exhibits.

base followed by an international manned space expedition to explore the planet Mars.

Leading and coordinating both the American and international efforts is the National Aeronautics and Space Administration (NASA), created in 1958 by the U.S. government. NASA's charter calls for the space agency to "plan and direct projects aimed at the peaceful exploration of space and . . . [gain] the scientific knowledge that will lead to betterment of the national way of life." Implementing the U.S. space program and cooperating with the programs of other international and national space organizations are NASA's 12 national space centers. Each center is dedicated to an important aspect of the only space program on Earth that has successfully landed humans on the moon and dispatched unmanned satellites to planets of our solar system and to the stars beyond.

NASA Headquarters

Directing and supervising the U.S. space program is NASA Headquarters. Conveniently located in Washington, D.C., NASA is conveniently located a very short walk from the National Air and Space Museum, where a living history of American aviation and space exploits is on display—the air- and spacecraft that conquered the skies and space. At the space agency's headquarters, some 1,500 specialists develop policies and guidelines for administering a multibillion-dollar budget that is spent on programs carried out by the space flight and research centers and other installations that constitute NASA. Among the major NASA offices in the nation's capital are the Office

Preceding page: This shot of a rocket booster shows the huge thruster nozzles that power it during the initial launch phase. *This page, top:* An intricate rocket engine underscores the complex design and packaging that was required for it to fit inside the body of a rocket booster. *Right:* Another rocket booster engine illustrates the complex plumbing necessary to burn off the powerful liquid oxygen at a preset combustion rate that provides the thrust needed to escape Earth's atmosphere.

Tourists line up, ready and waiting, to see a static display of spacecraft and boosters at the Kennedy Space Center. *Left:* An early model of a space shuttle, whose nose is composed of special tiles to withstand the thousands of degrees of heat during reentry to the Earth's atmosphere, fronts a field of rocket boosters.

of Aeronautics, Exploration, and Technology, the Office of Space Flight, the Office of Space Science and Applications, the Office of Space Operations, the Office of Commercial Programs, and the Office of Safety and Mission Quality.

John F. Kennedy Space Center

Perhaps the best-known American space center since the pioneer years of the *Mercury* and *Gemini* man-in-space missions and the *Apollo* lunar program, the John F. Kennedy Space Center (KSC) is located at Cape Canaveral on the east coast of Florida, approximately 50 miles east of Orlando. This center, known as Cape Canaveral in its earlier years, was renamed after the late president, who in 1961 decreed that NASA land Americans on the moon by the end of the decade. *Apollo* astronauts successfully carried out that mission in July 1969.

The Kennedy Space Center, 34 miles long and five to 10 miles wide, serves as the primary base for the checkout and launch of payloads and manned and unmanned space vehicles thanks to its location. KSC provides sufficient space and adequate safety to the surrounding civilian community during launches, landings, and other hazardous operations. Non-operational areas of the KSC complex have been declared a wildlife refuge.

Lyndon B. Johnson Space Center

"Houston, Tranquility Base here. The Eagle has landed." Those electrifying words reported by astronaut Neil A. Armstrong, the first person to step

Top: The space shuttle's immense size is evident when seen beside visitors at Kennedy. *Left:* Space shuttle displays at Kennedy's Visitors Center provide a narrative history of the development and flight of this space-age workhorse.

At the Kennedy Center's NASA press site, a space shuttle with boosters attached reaches for the sky. The space shuttle program is expected to continue through the end of this century and assist in the construction in space of the manned space station *Freedom*. *Below, Left:* A spacecraft nose cone is on display at Kennedy's Visitors Center. The nose cone protects the spacecraft during the launch phase, falling away once spaceborne. *Right:* Also called lunar landers, lunar modules were used to transport two astronauts to the moon's surface while a third astronaut orbited the moon in the command module.

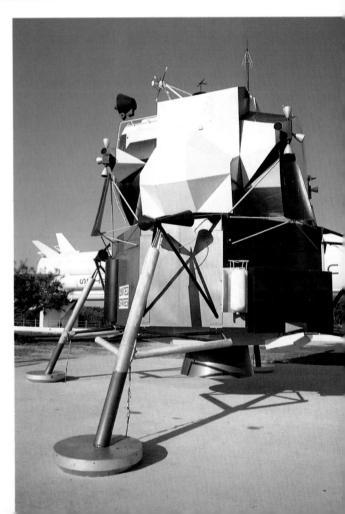

foot on the moon, on July 20, 1969, were heard first at the Lyndon B. Johnson Space Center (JSC), in Houston, Texas.

"Roger, Tranquility, we copy you on the ground," responded Mission Control Center (MCC), also known as Mission Control, in a live broadcast seen and heard by hundreds of millions of people in virtually all nations. The Johnson Space Center, named for the late president Lyndon B. Johnson in 1973, was originally NASA's Manned Spacecraft Center (MSC), which was established in 1961. This is NASA's primary center for the design, development, and testing of spacecraft; selection and training of astronauts; planning and conducting of manned missions; and extensive participation in the medical, engineering, and scientific experiments carried out aboard spacecraft.

Perhaps the most filmed and photographed building at Johnson, Mission Control is a three-story structure containing some of the world's most sophisticated communication, computer, and data reduction equipment, as well as wall-size data displays that lend a science-fiction aura to the two massive Flight Control Rooms (FCR), or "fickers," as they are called by the NASA staff. NASA's FCR is located on the second floor; the Pentagon's, used for both manned and unmanned secret space flights, is on the third.

Top: A young visitor stands before a space suit, used by an astronaut who conducted early space walks, outside the *Gemini* spacecraft. *Left:* The National Aeronautics and Space Administration (NASA) insignia can't be ignored at the Kennedy Space Center.

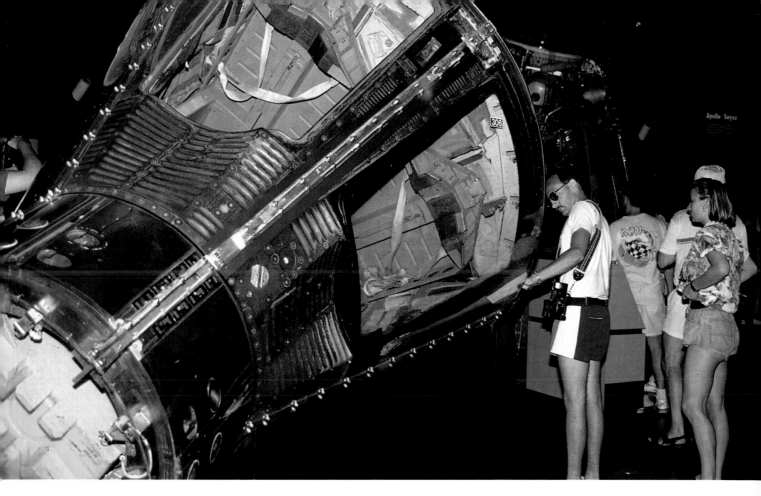

The two-man *Gemini* flights trained astronauts for future lunar missions and provided NASA engineers with the information required to design the *Apollo* spacecraft, the lunar landers, and the command modules. *Below:* The curvature of the Earth adds a startling backdrop for the *Gemini 7* spacecraft as it orbits during a rendezvous mission with *Gemini 6* on December 15, 1965. This photo was taken by astronauts in *Gemini 6*.

An astronaut takes a "space walk." Edward H. White II was the first American to take a space walk—which lasted 21 minutes—on June 3, 1965, during the flight of *Gemini 4*. He was secured by a 25-foot umbilical line and a 23-foot tether, and used a small hand-held compressed gas unit to maneuver.

The unmanned *Skylab 1*, launched from Kennedy on May 4, 1973, suffered some damage during the launch, which was later repaired by astronauts. The repairs made to the *Skylab* orbital workshop proved that astronauts could carry out many roles in space, including repairs to spacecraft. *Below:* The interior of the *Skylab 1* spacecraft is viewed by tourists at Kennedy's Visitors Center. *Following page:* Working outside the orbital workshop of *Skylab 3,* scientist/astronaut Owen K. Garriott deploys the particle collection experiment that is mounted on one of the solar panels to collect material from inter-planetary dust particles for study back on Earth.

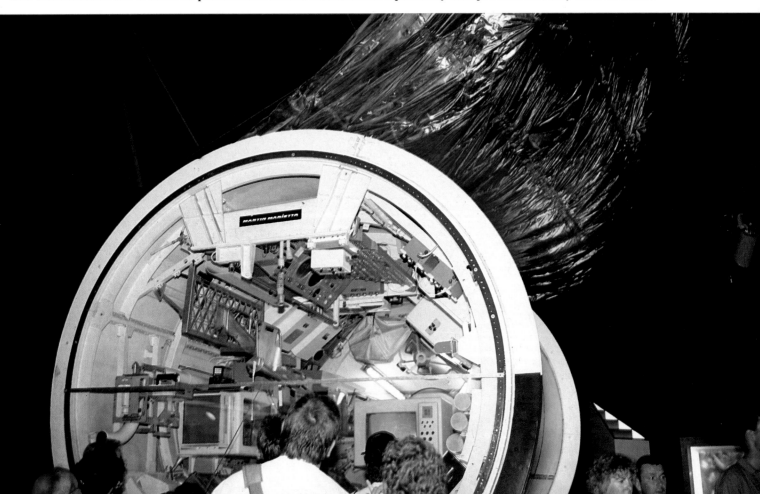

George C. Marshall Space Flight Center

Originally established as the U.S. Army Ballistic Missile Agency, this facility in Huntsville, Alabama, was staffed by military personnel and German space scientists rounded up after the collapse of Hitler's "Thousand Year Reich" at the end of World War II. It was named after the late World War II U.S. Army general and chairman of the U.S. Joint Chiefs of Staff, who later became Secretary of State. President Dwight D. Eisenhower dedicated the facility in September 1960 and turned it over to NASA. Located inside the U.S. Army's Redstone Arsenal, Marshall Space Flight Center (MSFC) became NASA's center for the development of launch vehicles and spacecraft.

The Marshall Space Center employs some 3,600 civil service personnel, of whom 58 percent are scientists, engineers, and technicians; the remainder are business professionals involved in developing and supervising Federal contracts for space programs to be carried out by aerospace companies. The MSFC is also a multiproject management, scientific, and engineering establishment centered around projects involving scientific investigation and application of space technology in solving Earth's problems.

The Huntsville facility had a significant role in the development of the Spacelab, the Hubble Space Telescope, and the Space Shuttle's powerful booster rockets and the management of the Advanced Solid Rocket Motor that will eventually

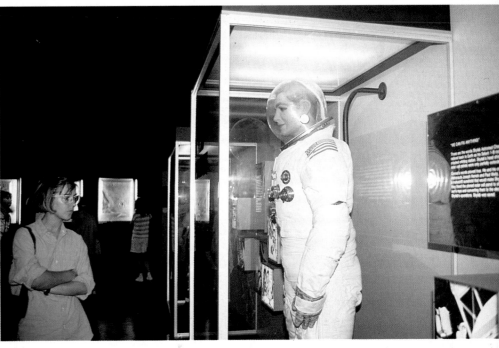

Top to bottom: An early model of the one-man *Mercury* spacecraft, originally called a "space capsule," is viewed by visitors at the Kennedy Space Center. A *Mercury* space suit is among the many objects on exhibit. A selection of space suit models hangs from a rack in the U.S. Air Force Space Museum at Kennedy.

Elements on the Moon

Visitors to Kennedy's Lunar exhibit study photographs of the moon taken by astronauts who landed there. *Below, left:* A detailed living history of the U.S. space effort is available to anyone who visits the Kennedy complex. *Right:* Samples of lunar rock and dust brought back to Earth by the *Apollo* astronauts include geologic examples from different parts of the moon.

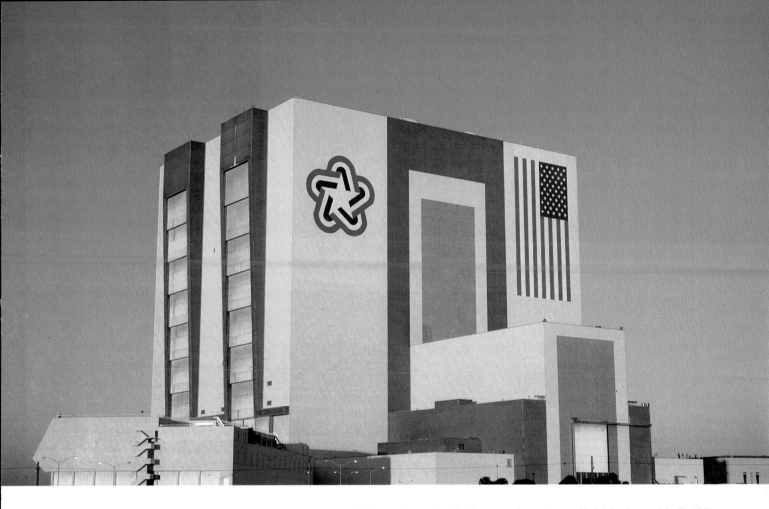

Booster rockets and spacecraft are assembled for delivery to the launch site in the huge and cavernous Vehicle Assembly Building at the Kennedy Space Center. *Below:* A view of the length of the Vehicle Assembly Building boggles the minds of visitors. Here rocket boosters, and spacecraft mounted above them, can top off upwards of 200 feet.

FROM THIS SITE, COMPLEX 14 AT CAPE CANAVERAL AIR FORCE STATION, ASTRONAUT JOHN GLENN, JR. BECAME THE FIRST AMERICAN TO ORBIT THE EARTH. HIS THREE-ORBIT FLIGHT ON FEB. 20, 1962 WAS FOLLOWED BY LONGER FLIGHTS BY SCOTT CARPENTER, WALTER SCHIRRA, AND GORDON COOPER TO COMPLETE PROJECT MERCURY, AMERICA'S FIRST STEPS ON THE PATH TO SPACE.

DONATED BY THE AMERICAN MONUMENT ASSOCIATION

replace liquid-fueled boosters. The Marshall Space Center also plays a key role in the development of payloads that are sent into space, some on missions to distant planets and stars.

Langley Research Center

Shortly after the United States entered World War I in 1917, the U.S. Army established the nation's first aeronautical laboratory at what is now the Langley Research Center (LRC) in Hampton, Virginia. Occupying 787 acres of government-owned land and sharing aircraft runways, utilities, and some facilities with the U.S. Air Force at nearby Langley Air Force Base, this center's prime mission, which comprises up to 60 percent of the projects at the facility, is basic research in aeronautics. NASA employees at LRC are dedicated to improving today's aircraft and to developing concepts and technology for the future ranging from general aviation and transport aircraft to the National Space-Plane hypersonic transports.

Langley's primary goal is to develop technologies that will enable aircraft to fly faster, farther, and higher and be safer, more maneuverable, quieter, more energy efficient, and less expensive to build. Over 40 wind tunnels and other unique research facilities and testing techniques backed by computer modeling capabilities cover the full flight range. An additional 3,200 acres of marshland is used by NASA as a drop zone for model aircraft tests.

Future hypersonic aircraft will require heat-resistant materials and supercomputers for engine and airframe design, and the fabrication of composite materials that will withstand thousands of degrees of heat

Top: A tablet at the site of the *Mercury* launch pad is dedicated to the flight of astronaut John Glenn, Jr., who was the first American to orbit the Earth on February 20, 1962. *Left:* An aerial view shows the huge aircraft landing strip at Cape Canaveral, which is used to return the space shuttle atop a special jet transport. In an emergency, it can be used as a landing strip by the space shuttle.

The underside of a space shuttle is being prepared in the Vehicle Assembly Building at the Kennedy Space Center. Once work is completed, the space shuttle is upended on a special rail car for transport to the launch pad. *Below:* The space shuttle *Discovery* will soon be transported from the Vehicle Assembly Building to the launch site.

Left: The countdown, already started for this space shuttle launch, will continue to the final 10-second launch sequence countdown that will end when the launch director signals, "We have lift-off." *Below:* A space shuttle, already prepared for pre-launch activities, is slowly transported to the launch site where it will be fueled for a mission and made ready for lift-off. *Opposite:* Final preparations are being made for the launch of this space shuttle, as technicians, working round-the-clock, double-check all instrumentation systems.

Atlantis heads for the high frontier following a textbook launch. The space shuttles, which can carry a crew of up to nine people, have been used to launch telecommunications satellites and to recover damaged satellites, which are returned to the manufacturer for repair. *Left:* In a blaze of smoke and fire, the space shuttle *Discovery* thunders into space in a perfect launch from the Kennedy Space Center.

WHENEVER MANKIND HAS SOUGHT TO CONQUER NEW FRONTIERS
THERE HAVE BEEN THOSE WHO HAVE GIVEN THEIR LIVES FOR THE CAUSE.
THIS ASTRONAUTS MEMORIAL, DEDICATED MAY 9, 1991, IS A TRIBUTE TO
AMERICAN MEN AND WOMEN WHO HAVE MADE THE ULTIMATE SACRIFICE
BELIEVING THE CONQUEST OF SPACE IS WORTH THE RISK OF LIFE.

THE ASTRONAUTS MEMORIAL FOUNDATION

The Challenger monument at the Kennedy Space Center was built as a memorial to the seven astronauts who lost their lives on January 29, 1986, as spectators and television viewers around the world watched Mission STS 51-L explode more than a minute after launch. *Below:* The fateful *Challenger's* crew consisted of astronauts Ronald E. McNair, Ellison S. Onizuka, Dr. Judith A. Resnik, Francis R. Scobie, and Michael J. Smith; payload specialist Gregory Jarvis; and schoolteacher Sharon C. McAuliffe.

FRANCIS DICK SCOBEE   RONALD E MCNAIR
MICHAEL J SMITH   S CHRISTA MCAULIFFE
ELLISON S ONIZUKA   GREGORY B JARVIS
JUDITH A RESNIK

THEODORE C FREEMAN

VIRGIL GUS GRISSOM
EDWARD H WHITE II
ROGER B CHAFFEE

generated by reentry into the Earth's atmosphere prior to landing. These hypersonic aircraft designs are for spacecraft that will take off like a supersonic jet transport; attain extreme altitude along the edge of the Earth's stratosphere where space begins; and travel at speeds ranging from Mach 1, the speed of sound (or nearly 800 miles per hour), to Mach 25, an Earth-orbiting speed of 19,000 mph.

The remainder of NASA's research at Langley centers on the development of technology for advanced space systems; research in laser energy conversion techniques for space applications; design studies for large space-systems technology, such as the space station *Freedom;* and the development of construction techniques, systems engineering and integration, and the design of large space antennae.

Goddard Space Flight Center and Wallops Flight Facility

One of the largest scientific and technical archives of space flight experiments is located at this sprawling campus 10 miles northeast of Washington, D.C. The Goddard Space Flight Center (GSFC), in Greenbelt, Maryland, consists of a multitalented space flight team of scientists, engineers, technicians, and project managers dedicated to extending the horizons of human knowledge, not only about the solar system and the universe, but also about the Earth and its environment.

Top: A rocket engine and booster make an impressive display at the Lyndon B. Johnson Space Center in Houston, Texas, known to a generation of Americans as "Mission Control," the central command post for all U.S. space flights once they are successfully launched. *Right:* Visitors examine a rocket booster engine housing at the Johnson Space Center. Established in 1961 to direct the *Apollo* lunar missions and formerly called NASA's Manned Spacecraft Center, in 1973 the center was renamed for President Lyndon B. Johnson, who strongly supported the U.S. space program.

This space-age mural is one of several at Johnson, NASA's center for design, development, and testing of spacecraft; selection and training of astronauts; planning and supervision of manned missions; and extensive participation in the medical engineering and scientific experiments carried out aboard spacecraft. *Below, left:* A display at Johnson features the electric-powered four-wheel lunar rover and other critical items used by *Apollo* astronauts to successfully carry out lunar missions. *Right:* The tiny lunar module (LM) was capable of landing just two *Apollo* astronauts on the moon; once it transported them back to the command module, it was set free and fell back to the moon.

A lunar module makes an actual descent to the moon's surface. Upon safely landing, the two astronauts aboard will unload the various items they will need to carry out the many tests and experiments they have been assigned.

A special suit worn by space shuttle astronauts who have to work in the open cargo bay is carefully examined by a NASA technician at the Johnson Space Center. *Below, left:* An umbilical line supplies air and telemetry to the astronaut working in the space environment; a second safety cord is attached to the astronaut and anchored in the spacecraft for added safety. *Right:* One of many various configurations for spacecraft is this mock-up, complete with dummy, of a space laboratory at Johnson.

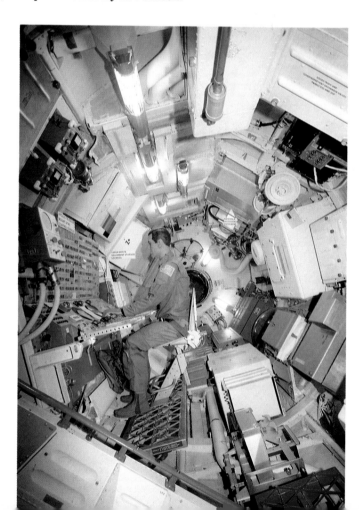

Named after Dr. Robert H. Goddard, the father of American rocketry, this center's research revolves around six space and Earth science laboratories and the management, development, and operation of several near-Earth space systems. At Goddard, the scientific and technical data from these major space flight experiments and programs are catalogued and archived at the National Space Science Data Center in the form of magnetic tapes, microfilm, and photos to satisfy the thousands of requests each year from the scientific community.

Once launched into space, manned and unmanned spacecraft are monitored around the clock by a worldwide ground and space-borne communications network, the nerve center of which is located at Goddard. One of the key elements of that network is the Tracking and Data Relay Satellite System (TDRSS) and associated ground tracking stations. One of the prime missions of TDRSS will be the relay of communications to and from the permanently manned space station *Freedom* once it is constructed in space. Goddard's role in the space-station program is the development of the telerobotic servicer that will be used to assemble and service operations at the station.

Among these major Goddard projects are the Cosmic Background Explorer (COBE), deployed in late 1989 to test the theory that the universe began about 15 billion years ago following a cataclysmic explosion—the "Big Bang" theory—and then expanded. Planned for the 1990's is the Gamma Ray Observatory (GRO), utilizing the Energetic Gamma-Ray Experiment Telescope

Top: The Anechoic Chamber is used to test materials to learn how well they stand up to reverberations and other high and low frequency sounds that affect astronauts as well as equipment. *Right:* Lunar rock brought back by *Apollo* astronauts is constantly being studied at the Johnson Space Center, which has one of the biggest collections of moon rock and soil.

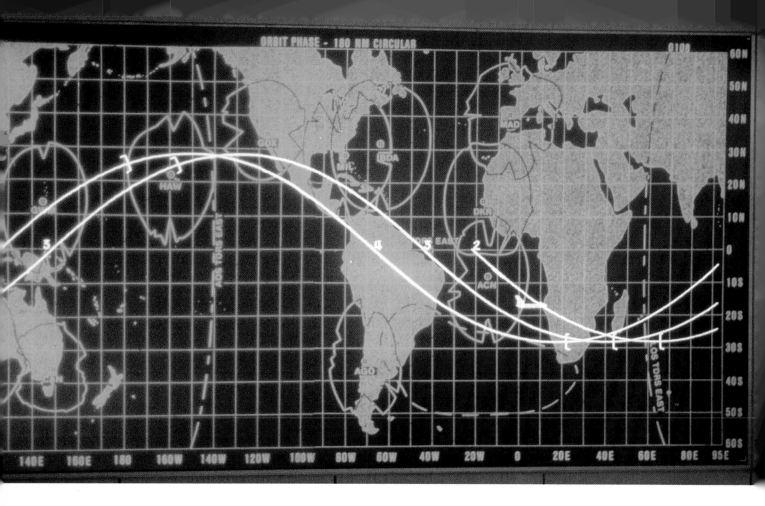

A large lighted map of the world in the Mission Control room depicts the orbital tracks of a spacecraft prior to its return to Earth. *Below:* Technicians monitor astronaut training at consoles at the Johnson Space Center. The electronic monitors provide trainers with complete information about an astronaut's performance from heartbeat to blood pressure to breathing rate: Nothing is left to chance.

Astronaut James B. Irwin, *Apollo 15* pilot of the lunar module *Falcon,* salutes the American flag that was planted at the moon's Hadley-Apennine landing site on August 1, 1971. *Following pages:* As the Earth spreads out below, an Agena Target Docking Vehicle is photographed from the *Gemini 10* spacecraft during a rendezvous mission; once they link up, the Agena's rocket is fired to boost the two joined spacecraft into a new orbit. Astronaut Charles M. Duke, Jr., pilot of *Apollo 16* lunar module *Orion,* stands on the moon beside the battery-powered lunar rover or "dune buggy," as it was called by the astronauts. It was Christmas Eve, December 24, 1968, when *Apollo 8* astronauts Frank Borman, James A. Lovell, Jr., and William A. Anders photographed this picture of the Earth while they orbited the moon. It was the very first voyage to the moon and the three Americans, inspired by the sight, read excerpts from the Book of Genesis.

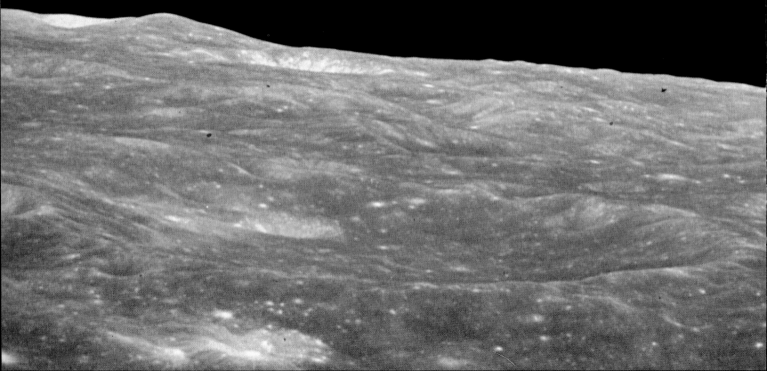

(EGRET) built by a Goddard laboratory. The GRO and EGRET are designed to penetrate the little-understood processes that propel the energy-emitting objects of deep space—the exploding galaxies, black holes, and quasars. The Upper Atmosphere Research Satellite (UARS), developed by Goddard, will look back at the Earth's atmosphere to help scientists understand its composition and dynamics.

Goddard's tracking responsibility also extends to its Wallops Flight Facility on Wallops Island, Virginia, one of the world's oldest launch sites. Established in 1945, this five-mile-long and half-mile-wide launch facility is located on Virginia's eastern shore. Wallops manages and directs NASA's sounding rocket program, scientific balloon program, Earth and ocean physics and biological and atmospheric science studies, and developmental projects for new remote sensor systems experiments. In addition to its approximately 100 rocket launches each year, the island supports Pentagon and other Federal agencies in aeronautical research with up to 200 aircraft and airport test operations annually at the research airstrip on the Virginia coast opposite the island.

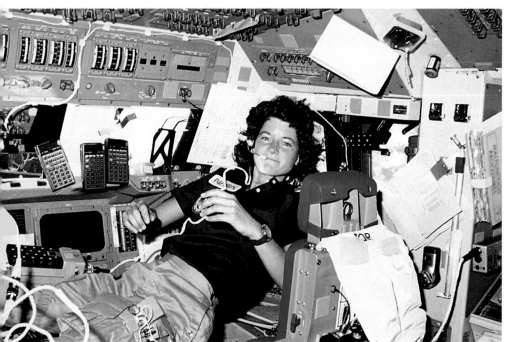

Jet Propulsion Laboratory

A history of the Jet Propulsion Laboratory (JPL) in Pasadena, California, embraces the U.S. space program from its inception after World War II to the present. Operated by the California Institute of Technology under a NASA contract, the laboratory manages the Deep Space Communications Complex, situated on 40,000 acres of U.S. Army land in Goldstone, California.

Top to bottom: Astronaut Richard H. Truly, space shuttle *Challenger* mission commander, talks to ground controllers from the flight deck on August 30, 1983. Astronaut Sally K. Ride, the first American woman in space, is comfortably seated aboard the *Challenger*. Astronaut Guion S. Bluford, Jr., the first African-American in space, maintains muscle tone on a treadmill aboard the *Challenger*.

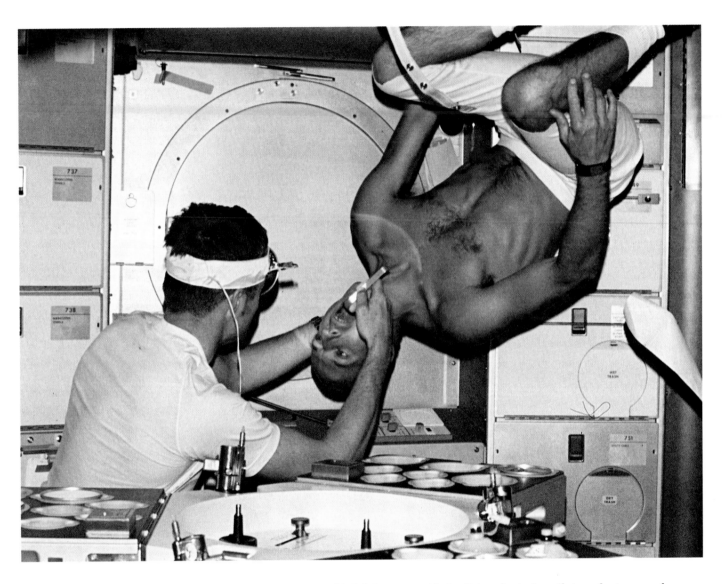

"Open wide and say ah...." Dr. Joseph P. Kerwin, *Skylab 2* astronaut/scientist and mission pilot conducts an oral examination of astronaut Charles Conrad, Jr., mission commander.

A space shuttle display is placed near the entrance to the George C. Marshall Space Flight Center in Huntsville, Alabama. Once the headquarters of the U.S. Army Ballistic Missile Agency, Marshall was named after the late U.S. Army general and joint chiefs chairman and secretary of state. *Below:* Visitors crowd around the aft section of a *Saturn V* rocket engine at Marshall; for many years the *Saturn V* was the most powerful single rocket booster in the U.S. inventory. *Opposite:* The foreground of this rocket display at Marshall is dominated by the Mercury-Redstone booster, which was developed to launch the U.S. man-in-space program embodied by the Project Mercury missions of the early and mid-1960s.

Marshall Space Flight Center's 3,600 employees, 58 percent of whom are scientists, engineers, and technicians, have become leaders in the development of launch vehicles and spacecraft. *Below:* Marshall features examples of some of the diverse launch vehicles and spacecraft it has developed. The center has had a significant role in the Spacelab and Space Shuttle programs, along with management of the Advanced Solid Rocket Motor that will one day replace liquid-fueled boosters.

The *Apollo 16* command module was designed at the Huntsville complex, which is also a multi-project management, scientific, and engineering establishment specializing in scientific investigation and application of space technology to help solve Earth's problems. *Below, left:* Impressive pictorials remind visitors of Marshall Space Flight Center's role in the national space program. *Right:* A tour guide sits in a lunar rover at Marshall—developed there for use during two *Apollo* landings on the Moon.

Marshall has produced a mock-up of part of the *Freedom* space station that will be constructed in space beginning in 1995 by a consortium of U.S. and foreign nations. *Below, left:* A model of a Russian spaceprobe is on display at the Marshall Space Flight Center. With the demise of the Soviet Union, the U.S. now expects more space cooperation with the former Soviet space program. *Right:* Among the Soviet space items on display at Marshall is this space suit worn by cosmonauts during their missions in space.

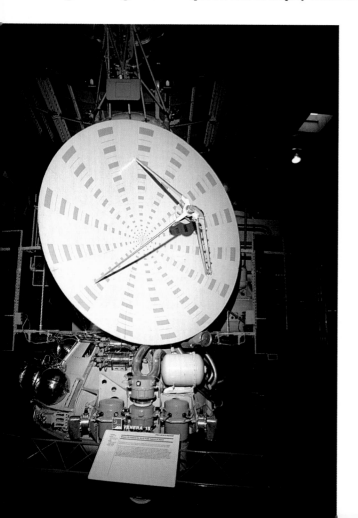

JPL is one of several stations that are part of the worldwide Deep Space Network (DSN) with similar tracking sites located around the world.

The "Lab," as it is known, is also engaged in the design and development of deep-space automated and unmanned scientific missions that have been and are carried out by Voyager, Galileo, Magellan, Mars Observer, Comet Rendezvous Asteroid Flyby, and a host of other space exploration satellites.

John C. Stennis Space Center

The John C. Stennis Space Center (SSC), in Bay St. Louis, Mississippi, is NASA's premier center for testing large rocket propulsion systems for the Space Shuttle and future generations of space vehicles. Originally the National Space Technologies Laboratory when it was formed in 1963, it was renamed after the late Mississippi senator who supported the nation's space program.

Ideally located on Mississippi's East Pearl River, Stennis Space Center's 13,480-acre operations complex includes industrial and specialized engineering laboratories and test beds to support engine testing and the transport of huge boosters, oversize cargo, and liquid rocket-fuel propellants aboard deepwater barges that are towed along the Intracoastal Waterway. The SSC is surrounded by an additional 125,828 acres that act as an acoustical buffer zone to muffle the loud noise produced during static engine tests.

In addition to a laboratory for ground testing, SSC has evolved into a center of excellence in the area of

Top: A space camp for youngsters, who train on simulators like the astronauts, is located at the U.S. Space & Rocket Center in Huntsville alongside NASA's Marshall Space Flight Center. A year-round program offering young people different levels of participation has drawn visitors from around the world. *Right:* Astronauts also train underwater, and this scuba training tank at the U.S. Space & Rocket Center accommodates youngsters who have reached Level 2 training at the unique camp.

Critical materials testing is carried out in this X-ray astrophysics facility at the Marshall Space Flight Center. Constant specialized testing of materials and equipment is a must if they're to be used in the space environment. *Below:* Technicians work on the *Freedom* space station mock-up. The Huntsville engineers and scientists leave nothing to chance in their pursuit of excellence to further the U.S. space program.

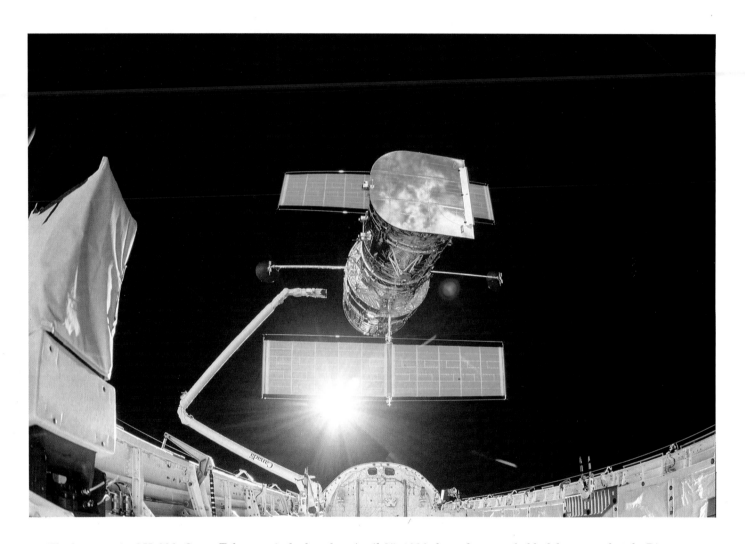

The long-awaited Hubble Space Telescope is deployed on April 25, 1990, from the cargo hold of the space shuttle *Discovery*. With 15 years in the planning and construction, this telescope is expected to open new horizons for astronomers.

One of the world's largest wind tunnels for aircraft flight testing is at NASA's Langley Research Center in Hampton, Virginia. It is kept busy by U.S. government agencies and air transport industries responsible for the global leadership and safety of America's aircraft. *Below:* A full-scale aircraft undergoes wind tunnel testing at Langley, whose primary goal is to improve today's aircraft and develop technologies and concepts that will permit them to fly faster, farther, higher, and safer. Langley also develops future aircraft for both general and commercial aviation.

An Agena Target Docking Vehicle is displayed at the Robert H. Goddard Space Flight Center in Greenbelt, Maryland. *Below:* Satellite and sounding rocket experiments performed at Goddard provide important information to scientists about the Earth's environment, its relationship to the sun, and the universe.

An array of antennae located on the 40,000 acres of the Jet Propulsion Laboratory (JPL), in Pasadena, California, probes deep space for some of the automated, unmanned scientific satellites developed at the "Lab." *Below, left:* A visitor studies the planet mapping globe at JPL's Space Museum. Operated for NASA by the California Institute of Technology, JPL manages the Deep Space Complex and has designed and developed a host of unmanned space exploration satellites such as Voyager, Galileo, Magellan, and Mars Observer. *Right:* The JPL engineers who designed and developed the Galileo unmanned deep-space probe, similar to this model in the JPL Space Museum, solved a problem aboard the $1.4-billion satellite across 146 million miles of space. Galileo is expected to fly by Jupiter in December 1995 — on schedule.

remote sensing with projects involving both national and international Earth sciences programs. NASA has designated SSC and its approximately 4,500 employees as the space agency's primary center for commercialization of remote sensing from space.

Lewis Research Center

NASA's Lewis Research Center (LRC), in Cleveland, Ohio, began as a U.S. government aircraft-engine research center in 1941. Lewis conducts research and development of aircraft propulsion, along with space propulsion and space power and satellite communications. Located in 100 buildings spread across 360 acres adjacent to Cleveland's Hopkins International Airport some 20 miles southwest of Ohio's largest city, LRC has been advancing propulsion technology that will enable aircraft to fly faster, farther, and higher, using less jet fuel; noise and pollution would also be reduced.

Lewis has in addition been a pioneer in the use of high-energy fuels for both air breathing and space propulsion by developing liquid hydrogen as a rocket booster fuel. Among this center's other responsibilities are the development of the largest space power system ever designed to provide the necessary electricity to accommodate life support systems and research experiments that will be conducted aboard *Freedom*.

Space Telescope Science Institute

One of the newer NASA facilities, the Space Telescope Science Institute, on the Johns Hopkins Homewood Campus in Baltimore,

Top: The Hubble Space Telescope operated for NASA by the Space Telescope Science Institute of Johns Hopkins University, in Baltimore, Maryland, slowly rises from the cargo hold of the space shuttle *Discovery* in April 1990. *Right:* Prior to its launch, the Hubble Space Telescope is prepared for delivery to the cargo hold of the space shuttle. Engineers from the U.S. and Europe control the telescope from the institute's command center.

Among the new horizons that the Space Telescope Science Institute (ST ScI) is expected to open for astronomers using the Hubble Space Telescope will be new information about the planet Saturn and its mysterious rings. *Below:* The institute will also focus the telescope on deep space phenomena like this mysterious supernova ring photographed in a nearby galaxy thousands of light years away. *Opposite:* The space shuttle *Challenger* blasts off on August 30, 1983 with five astronauts aboard and a cargo hold filled with various scientific experiments.

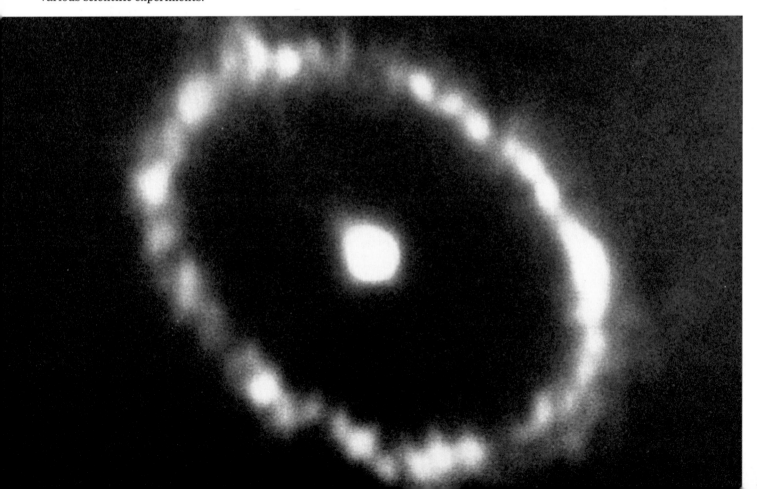

Preceding page: A space shuttle that has landed at NASA's Ames-Dryden Flight Research Facility at Edwards Air Force Base, in California's Mojave Desert, is parked under the lights at the sprawling facility while technicians prepare it for transportation back to the Kennedy Space Center. *This page, above:* The space shuttle *Enterprise* on a test flight prepares to land on the dry, hard-packed Mojave Desert floor at Edwards Air Force Base some 80 miles north of Los Angeles. *Below:* Awaiting a piggyback flight aboard a NASA 747 jet transport, the space shuttle *Columbia* will be flown from the Ames-Dryden Flight Research Facility across the country to the Kennedy Space Center in Florida.

A security guard stands in the test section of the wind tunnel, a 40-foot wide and 80-foot long area within the 120-foot long facility at the Ames-Dryden Flight Research Facility, which is dedicated to testing manned and unmanned aircraft.

Maryland, is operated by the Association of Universities for Research in Astronomy and includes resident scientists and engineers from the European Space Agency. Bearing the acronym ST ScI, this campus facility conducts science operations for the Hubble Space Telescope, and computer and imaging systems of radio telescopes and other remote sensors aboard deep-space-probe satellites.

Hugh L. Dryden Flight Research Facility

NASA's Dryden Flight Research Facility, in Edwards, California, is located in the Mojave Desert some 80 miles north of Los Angeles. Isolated but adjacent to Edwards Air Force Base, the facility enjoys almost ideal weather for flight testing and landing at nearby Rogers Dry Lake, a 44-square-mile natural surface for landing the space shuttles and other test aircraft.

Employing some 450 civil service personnel and nearly 400 contractors, the center's primary research since 1947 has been dedicated to testing manned and unmanned aircraft in an area free from population disturbances or natural hazards. Ground-based facilities include a high-temperature loads calibration laboratory for testing complete aircraft; a highly developed aircraft flight instrumentation capability; a flight systems laboratory; and an elite organization of test pilots, engineers, technicians, and mechanics unmatched anywhere.

Top to bottom: Hanging from the high ceiling of the National Air and Space Museum in Washington, D.C. are actual aircraft of the 1930s – a biplane, a U.S. Army Air Corps P-36 pursuit plane, and the P-40 that was used in 1939-41 by American Volunteer Group pilots who flew for China against the Japanese. Aircraft used during the years from World War I through World War II are on display at the National Air and Space Museum, which is located a very short walk from NASA headquarters. Among the aircraft are an early version of the Douglas DC-3 twin-engine airline transport. A full-size replica of the Wright Brothers' aircraft *Kitty Hawk*, which made the first powered flight on December 17, 1903, hangs alongside a modern rocket booster at the National Air and Space Museum.

Preceding page: A huge rocket booster towers over the entrance to the Space Center in Alamogordo, New Mexico, one of many museums devoted to the exhibition of space-related artifacts. *This page, above:* A satellite hangs from the ceiling of the Space Center in Alamogordo. *Below:* Another exhibit is the interior of an RL-10 rocket engine.

This versatile center has not only demonstrated its capability with high-speed research aircraft, but also with such unusual flight vehicles as the Lunar Landing Module and wingless lifting bodies; Dryden is also responsible for the development of the National Aero-Space Plane and a hypersonic transport designed to fly above the Earth's stratosphere from New York to Tokyo in less than three hours.

Ames Research Center

Founded in 1939 as an aircraft research laboratory and named after Dr. Joseph C. Ames, Chairman of the National Advisory Committee for Aeronautics (NACA), the Ames Research Center, in Moffett Field, California, is located in the heart of Silicon Valley at the southern end of the San Francisco Bay. NASA merged the Ames and Dryden research centers in 1981 but kept Ames focused on scientific research, exploration, and applications directed toward creating new technology for the United States.

This space center's 1,750 civil service personnel and some 1,700 contractors, also supported by approximately 400 graduate students, post-doctoral fellows, and university faculty, concentrate on computer science and applications, experimental aerodynamics, flight simulation and research, hypersonic aircraft such as the National Aero-Space Plane, aeronautical and space human factors, life sciences, and solar-system exploration.

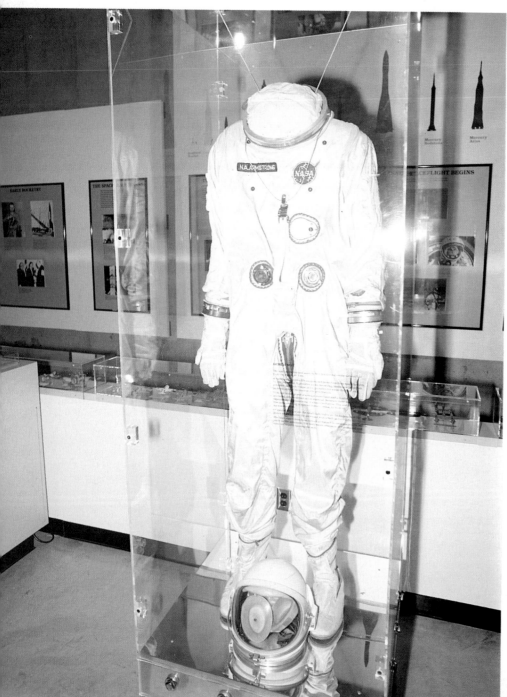

Top: An F5D Skylancer fighter is displayed at the entrance to the Neil Armstrong Air and Space Museum in Wapakoneta, Ohio, named after the astronaut who was the first man to walk on the moon. *Left:* A Project Gemini space suit, worn by astronaut Neil Armstrong when he flew his first space mission in a two-man *Gemini* spacecraft on March 16, 1966, is displayed at the museum.

A U.S. Navy jet stands on display at the Aerospace Historical Center in Jackson, Michigan, one of several cities that have established aviation and space museums. *Below:* Space-age boosters and rocket engines are displayed at the Balboa Park Aero-Space Museum in San Diego, California.

A selection of Strategic Air Command intercontinental ballistic missiles are at this U.S. Air Force Command Air Museum, Offutt AFB in Omaha, Nebraska. *Below:* A space suit and *Apollo* command module are framed in this display of space-age equipment at the Jackson, Michigan Space Museum. *Opposite:* A selection of rocket boosters and rocket engines used in the U.S. space program can be seen at the Museum of Science and Industry in Chicago, Illinois.

Index of Photography

All photographs courtesy of The Image Bank except where indicated *.